Emma Hardinge Britten

The electric physician

Self cure through electricity

Emma Hardinge Britten

The electric physician
Self cure through electricity

ISBN/EAN: 9783337210014

Printed in Europe, USA, Canada, Australia, Japan

Cover: Foto ©berggeist007 / pixelio.de

More available books at **www.hansebooks.com**

THE

ELECTRIC PHYSICIAN:

OR

Self Cure through Electricity.

A Plain Guide to the Use of Electricity, with
Accurate Directions for the Treatment
and Cure of Various Diseases,
Chronic and Acute.

BY

EMMA HARDINGE BRITTEN,

ELECTRIC PHYSICIAN,

TEACHER OF THE FRENCH AND VIENNESE SYS-
TEMS OF MEDICAL ELECTRICITY, LECTURER
ON ANATOMY, PHYSIOLOGY, AND PSY-
COLOGY, AND AUTHOR OF VARIOUS
WORKS ON PATHOLOGY, THE
VITAL FORCES, &C., &C.

———

PUBLISHED BY DR. WILLIAM BRITTEN,
AT 155 WEST BROOKLINE STREET, BOSTON, MASS.
AND TO BE HAD OF ALL BOOKSELLERS.

———

INTRODUCTION.

THE object of this little work is to supply to the heads of families and non-professional persons, the means of applying for themselves the best method of cure yet discovered, and to suggest to professional students and medical practitioners, how they can acquire the elements of a beneficent and invaluable remedial art, which at present forms no part of a collegiate education.

Many learned treatises on electricity are already before the world, but few if any purely practical methods have as yet been published which would enable the unscientific reader to apply electricity to the cure of disease on safe, reliable and scientific principles.

The essays put forth under the attractive guise of information on the uses of electricity as a remedial agent consist for the most part of details concerning the history of discoveries and discoverers, experiments and experimenters, theories which the experiences of every fresh decade of time will be sure to overthrow, and abstractions which can serve no other possible use than to fill up so many pages of a *salable* volume.

When the author has been questioned by her friends, patients or pupils, what work they could consult that would enable them to utilize electricity as a reliable remedial agent in family practice, she has felt keenly the necessity for some treatise or guide devoted solely to this purpose, and free from the mass of irrelevant matter with which the literature of this subject is overladen, and it is with a view to meet this much needed desideratum that the following unpretentious pages have been written.

The author does not deem it necessary to inform the reader how or when the uses of electricity were first discovered. No learned disquisition on the origin of its nomenclature is required to instruct a person suffering from a raging toothache how they can utilize electricity for its relief, nor need the hapless victim of a severe neuralgic attack, study out the names of all the great or little personages who have experimented with electricity, before they can apply it for their own benefit.

Neither does it seem fitting to preface these pages, or subtract one iota from their singleness of purpose, by introducing the elements of anatomy and physiology, a course so universally adopted by medical writers however special their subject may be, that we can now scarcely pick up from our area steps a printed handbill inviting sufferers of all classes to go and be cured by some universal nostrum without being treated to a full and comprehensive detail of the human system in all its tissues, organs and parts.

Whilst the labors which a Wilson or a Carpenter have spread over six hundred pages of a medical treatise, can thus be crowded up into one sheet of *popular literature*, whilst every book-stall abounds with anatomical knowledge "for the million," and every common school affords to the rising generation ample means of studying the generalities of physiological law, it would be a reproach to the intelligence of the age to suppose that any readers of this work can be wholly ignorant of such important subjects. Assuming then that those whom we address are at least fairly educated persons, that being so, they must be familiar with the general principles of anatomical structure and physiological action, and that their chief aim in the perusal of these pages is to learn what electricity can do for them in the way of alleviating suffering, and supplementing the imperfect text-books of so-called modern "medical science," we shall at once proceed to give plain and comprehensible directions for the application of electricity as the latest discovered and best method of remedial art now known to man.

It is only necessary to add, that the system practiced with unbounded success by the author in her own professional career, the methods of which are in part laid down here, are neither originated by herself, nor yet by any others who with egotistical pride announce themselves as the "sole discoverers of a system." The author has carefully and industriously observed the modes of electrical application practiced in many countries, and from some of the most eminent practitioners in France and Vienna she has gleaned hints which she has studied to improve upon. Several other methods, now common in America, but not so well known in Europe, have also furnished her with the grounds for fresh experiments, and these, pursued with earnest care and cautious observation, have enabled her to acheive the most satisfactory results in her own extensive practice, and justify her in offering to the public the simple and comprehensive rules contained in these pages.

Whilst the domestic practice herein recommended cannot be expected to meet crisises of imminent danger, or supersede the necessity of professional skill in cases requiring the attendance of an accomplished surgeon, a faithful study of this little work will enable any person of ordinary intelligence to treat successfully all the diseases herein prescribed for, and many more of a cognate character.

Faith in the indiscriminate use of electricity as a force, the crude idea that any benefit can be derived from the passage of a simple current through the body, or the supposition that some mysterious results of a healthful character can grow out of the hap-hazard administration of what used to be vulgarly denominated "shocks," are all opinions which the growth of intellectual knowledge tends rapidly to dissipate. Still there is far too little attention given to the scientific method of applying electricity as a curative art. Whilst the tension, quantity and direction of the current is stringently considered in the performance of those curious experiments by which the accomplished electrician aims at amusing a promiscuous audience, no special attention to such points seems to have been demanded for the introduction into the human system of a force that can kill as well as cure.

The laws of polarity which govern all the motions of the human frame, the action of the great nerve centres upon distant parts, and the nice adjustment of the force evolved from the battery with that which vitalizes the body, but, above all, the direction of the current to the nervous fibres that have to be traversed, together with a keen appreciation of the results to be expected, in a word, a thorough knowledge of how and where to introduce electricity into the living subject, with a clear understanding of what it will or will not do, seem to be branches of knowledge which have hitherto formed no part of the physician's education, and which the public have not had sufficient understanding of the subject to demand from all who pretend to practice, or direct the practice, of medical electricity.

Whilst the author proposes to teach in detail to her scholars such results as she has herself arrived at in all these directions, she deems it superfluous to enlarge upon them more at length in this place. They form the science of. medical electricity, and it is enough to insist on the necessity of studying that science and acting upon its guidance ere medical electricity can be safely or effectively administered.

For the benefit of those who may not have time or capacity to study the *modus operandi* of a scientific method for themselves, these pages are written, and however short they may fall of the full importance which belongs to the subject, they will at least carry the careful student safely and successfully through a new field of curative art and one where every footprint should be marked with that skill and exactitude which experience alone çan confer.

THE ELECTRIC PHYSICIAN.

SECTION I.   ON ELECTRICAL MACHINES.

It cannot be questioned that some batteries or electrical machines are better suited to medical purposes than others, just as some are more adapted to the use of the mechanic or telegraphist.

Electro-magnetism is a form of force which seems to be more in assimilation with the human organism than galvanism; hence we start with the proposition that a well-constructed electro-magnetic machine is of all others the most suitable that can be used for medical purposes.

The vulgar idea that electricity is equally effective as a remedial agent however it may be evolved, or introduced into the human system, is fast becoming exploded, and it is

11

now understood that certain modifications and conditions of electrical force are as essential to success in the cure of disease, as in the production of telegraphic signals.

It would be as tedious as unnecessary to enumerate the different forms of machinery that have been invented and are now in use for medical purposes.

Each manufacturer sets up a claim for peculiar excellence in some special machine, the true value of which, however, can only be decided by experience. In the application of electricity to the cure of disease, the first great desideratum is the evolution of such a quality of force as shall correspond, as nearly as possible, to the imponderable currents of the physical system, vaguely termed "the life principle." It is in this respect that experience declares in favor of electro-magnetism. The quality of force evolved in the electro-magnetic battery seems to be more agreeable, harmonious, and effective than that produced by other machines, and the wide range of disease in which its action (under the manip-

ulation of skilful operators) has been suc-
cessful, would almost induce the belief that it
may eventually be recognized as the universal
cure-all for every disease to which flesh is
heir. As to the machine itself, the special-
ties required of it are, first; a wide range of
power, susceptible of being increased or diminished at pleasure, from a minimum de-
gree of sensational fineness, to the utmost
maximum of strength the patient can comfortably bear. The next desideratum is
smoothness and equality in the current.
No spasmodic jerks, sudden stoppages, or
violent impulses should occur to distress the
subject, and impede the comfort of the application. Unless required for some peculiar
case, the administration of "shocks" should
be carefully avoided and the strength or fineness of the current should be entirely under
the control of the operator.

To obtain these results, together with all
the details necessary to ensure success in the
application of electricity, the utmost care
must be used in the construction of the

machine employed. Due consideration must be had in the adjustment of the magnet, the length, strength, and arrangement of the coil ; the fashion of the plates, screws, and nature of the elements used ; the quality of the conducting strings, together with the size and shape of the electrodes.

All these items are of as much importance in the administration of medical electricity, as the points of contact in the subject to which the current is applied, and it is often found that want of due care and appreciation of these details, traverses even the most skilful operator in his efforts at success.

From many years of experience and observation on the wonderful curative properties of medical electricty, the author is fully convinced, it only needs a full and correct knowledge of the precise effect which the force will produce upon every organ and tissue of the human body, together with a machine susceptible of accurate manipulation in the operator's hands, to effect a complete revolution in medical science, and induce suffering humanity to accept of electricity as nature's true and only restorative in all conditions of disease. Let it be distinctly understood however, that the author makes no claim in her own experience to possess this wide and accurate range of knowledge, nor does she believe that any but the profound egotist, or the irresponsible charlatan, will advance such claims in the present imperfect state of our electrical knowledge.

As a science, the understanding of electricity is yet but in its infancy. As a remedial agent its application is but just beginning to emerge from the realm of exeriment to that of method.

Whilst the value of electricity has long been very generally admitted as a cure for a certain class of nervous disorders, it is comparatively a new idea, and one which has only obtained within the last few years, to believe that its effects might be harmonized into a purely natural system of treatment for every form of human suffering. Proceeding with the utmost caution then, and all the reticence which modesty demands, to prescribe the use of electricity according to already well-proved experience, we must again reiterate as an absolute pre-requisite for success, the necessity of practicing with such a machine as will produce the proper quantity and quality of force desired, and can be governed with ease and certainty.

For domestic purposes, general use, and especially with a view of following out the instructions given in this guide, the most available machine that can be procured is the " Home Battery," manufactured under the express supervision of Dr. William Britten of Boston, and designed to meet the great and increasing demand for a cheap, reliable, and perfectly manageable method of applying electricity.

This excellent machine is not only simple in working, smooth and gentle in its action, but sufficiently varied in strength, and powerful in full force, to meet all the demands of ordinary practice ; its price, moreover, is so moderate as to place it within the reach of the most limited means, and no family that values the importance of health and the incalculable advantage of self-cure in disease, need be, or should be without it.

A full description of this most desirable instrument, with the address of the manufacturer, and all details as to price, &c., will be found at the end of this work, and to the advertisement in question the reader is referred for farther information.

SEC. II. THE ELECTRODES.

Every manufacturer of electric or galvanic instruments sends out with his machines a pair of strings and a pair of tin handles, or small cylinders, by way of electrodes. The strings are long conducting wires covered with silk, worsted, or rubber, terminating at each end in small metal pins, one of which is connected with the battery by means of two upright brass posts with holes for the reception of the pins, and screws to secure them, whilst the other end is attached to the electrode.

2

When we use a sponge-cup, the tin handles or any special form of instrument, the metal pin is secured into an appropriate hole and screw adapted to receive it, but in following out the directions herein laid down, the second pin will be more frequently attached to a set of plates which in the French and Viennese systems are used as electrodes. The attachments to these plates are made by inserting the metal pin into a small groove or slide fitted on to the centre of each plate, as will presently be described.

It used to be deemed sufficient to charge the subject with electricity by holding the tin electrodes in either hand, or else the force was introduced into the body through a sponge pressed into one of the electrodes, or water into which the handle was dropped. It would be difficult now to enumerate all the various and ingenious instruments that have been devised for electrical applications.

For the professional operator, especially for one who undertakes to treat surgical cases, many different instruments and special forms of electrodes are needed, but for merely general use, the only electrodes necessary to complete an entire electrical apparatus are the strings and tin handles furnished with the battery, a sponge-cup fashioned with a long handle and wooden bowl at one end, into which a fine soft sponge is pushed, and a set of plates.

The conducting strings sent out with the "Home Battery" will be found to have at one end a copper, and at the other a brass, pin. This difference in the metals is deemed of much importance by the manufacturer, and it is essential to observe that the brass pins should be connected with the battery, and the copper with the electrodes. From this rule there must be no departure, and the difference in the metels will be easily detected by the superior malleability of the copper, even though the shape and general appearance of the pins may not otherwise indicate their variety.

The set of plates so often alluded to in this guide consist of five differently sized pieces of thin sheet brass, neatly turned down and carefully smoothed round the edges, and fitted in the middle with a small half-inch slide or groove fastened to the plate with little rivets, and only raised enough above the surface of the plate to admit of the insertion of the copper pin at one end of the string. In sliding this pin into the groove (which should fit it snugly) it is necessary to turn the string back against the slide so as to prevent its being pulled out or displaced when used on the subject's body. The dimensions of these plates are as follows: No. 1 is a nearly square plate ten inches long by eight wide. This is used to sit upon, place under the feet, or lay over a large surface. No. 2 is eight inches long by four wide; generally adapted to lay across the abdomen, under the calves of the legs, or across other parts less extended than that occupied by No. 1. No. 3 is called the kidney plate from being frequently used to place across the loins and above the kidneys. The dimensions are seven and a half inches long and three wide. No.

4 is useful to place between the shoulders or on the sternum or breast-bone. Its length is six inches ; width, one and three-fourths. No. 5 is especially designed to apply at the nape of the neck and nearly fits that part, being four and a half inches long, and one and a half wide. These dimensions are given for the benefit of those who desire to procure their plates at home, or do not wish to send to Dr. Britten for them. The corners should be carefully rounded, and the edges, as before stated, neatly turned down and thoroughly flattened.

The sponge-cup, designed for rubbing over surfaces where a soft electrode is needed, together with instruments suitable for the eye, ear, and throat, can all be procured from any electrical or surgical instrument maker. Dr. Britten also supplies them when ordered. The sponge and eye-cups are absolutely essential for the practice directed in the guide, especially the sponge-cup, but these are the only electrodes necessary for domestic use.

It must be observed here, that all applications of electro-magnetism should be made with moist electrodes. For this reason the author has adopted the method of using bags made of cotton cloth sufficiently large to admit the plates without exposing any of the metal to contact with the skin. These bags being dipped into hot water, the plates (with the string pin inserted into the slide and carefully turned back, to secure the string from being pulled out), is then placed in the bag and laid with the smooth side against the skin, and the string and slide uppermost on the patient's body at whatever point may be desired. Over the upper side of the plate is then laid a piece of rubber cloth, the design of which is to protect the patient's garments from wet and cold by coming in contact with the damp bag. Thus the connection is effected between the patient and the battery. The two brass pins are placed at the positive and negative posts of the battery ; the strings connecting these with the copper pins at the other end are then connected by means of slides with the plates, and these are laid as above discribed in damp bags (one fold of wet cloth intervening between the skin and the metal), and the whole protected by pieces of rubber cloth, completes the arrangement by which the force is introduced into the system.

The tin electrodes, or any given form of instrument, may be substituted for the plates, but these are not used moist. If the sponge-cup is used, the sponge must be soft, fine, and even, and moistened like the bags. The water used on all occasions must be warm, so as not to chill the patient in the first application. It should also be observed that when the plates are arranged as desired, the patient should be carefully covered and protected against all chance of cold or chilliness. It is scarcely necessary to add, that the most scrupulous cleanliness should be observed in the use of the plates, bags, rubber cloths, sponges, &c., and that the battery itself should be kept free from dust or impurity.

Full directions are generally pasted into the covers of the machines sent out on sale, and this is so fully carried out in the Home Battery, that it would be needless here to add any farther instructions concerning the mechanism.

It is only necessary to add that the sulphuric acid used in the jar should always be *chemically pure.* It may run for about a month without changing unless it is found that the strength of the battery fails ; it is then most advisable to throw away the solution and make fresh, also to see to it that no salts accumulate on the plates, that no dust collects about the magnet, that the strings are in good order, and the screws all sufficiently tight.

With these precautions, there will be little else to contend with, than the ordinary wear and tear of the instrument and elements at work.

Sec. III. Of the Electric Circuit.

In these days of general intelligence it
seems almost superfluous to inform the reader
that the electro-magnetic current forms a circuit running from positive to negative, and on
again until arrested, and that the philosophy
of introducing electricity into the human
body is to permit the subject to intercept the
flow between the entering and exit portions
of the machine. To pass a current through
the human body, however, it is not sufficient
to have only one point of contact with the
battery. The current must be introduced
into the system by an electrode connected
with the positive pole and carried off with
equal precision by another electrode connecting the subject with the negative pole of the
battery. In the shocks administered by the
Leyden jar, the force is concentrated and dis-
charged at once, but in the electro-magnetic
battery a continuance of the flow is kept up,
which forms the circuit through which the
subject is permitted to intercept it, becoming

a part of that circuit as above described. Now we enter upon this very simple explanation, not supposing it will convey any new idea to nine persons out of ten, but as we happen more than once to have encountered those tenth persons, and have found that they were entirely ignorant of what an electric circuit meant, in a word, that they supposed the mere act of holding one electrode, or even of coming into contact with a battery in action, would be sufficient to produce "a shock," and convey electricity into the system, so we ask permission of the nine well-informed readers to explain, for the sake of *the tenth*, on what conditions alone the benefits of electricity can be obtained.

Most young persons, aye and elderly ones too, will remember in their school days the experiment of a number of persons joining hands whilst two at the extreme ends of the chain held the positive and negative electrodes which connected the whole number with a galvanic battery. As long as the hands of all the party were thus linked to the machine the current evolved passed through each one alike, but let a single hand be disjoined, thus breaking the connection between the chain whose ends were attached to the battery, and all sensation would instantly cease. The number of persons might be diminished or increased at will ; the two only might be left who held the electrodes, and these, if joining hands, would re-form the circuit. Let but one remain, and that one hold *both* electrodes, and still the circuit would be complete. In short, whether it be one or more bodies that intercept the flow, so long as the inter-rupting mass be connected with the two poles of the battery, it will receive all the force of the current evolved ; and it needs only to recall this simple and very familiar experiment to apprehend at once how the subject or patient becomes charged by being placed *en rapport* with the poles of a battery. To those who have even superficially studied the laws of electricity, but especially to such as understand what this subtle force does to the human body, and how it is transmitted from the nerve centres along their fibres as the conducting wires to the points desired, it will at once become apparent that the utmost care is demanded in placing the electrodes, energizing certain parts with the positive pole, or diminishing the excess of inflammatory action with the negative, as the nature of the disease and the character of the organs treated may require. In fact, herein consists the science of administering electricity as a remedial agent, and it is to this point that we shall now direct the reader's attention.

Sec. IV. The Rationale of Health, Disease, and Cure:

It should be remembered that the human body is itself vitalized by a force which, if not actually electricity, is of a sufficiently analogous character to justify our regarding man as a grand magnet with numerous vital and nerve centres, points of polarity, and circuits of motion, acting and re-acting in the mysterious processes of life, much on the principle

of machinery propelled by electricity. Disease itself is merely a disturbance of those
imponderable forces by which the integrity
of the human system is maintained, and consists either of an excess or deficiency of vital
action, manifest in the form of inflammation or
torpidity, and it is on this principle that we
deem the best restorative for disease must be
an application of the very force whose ana-
logue has been disturbed.

The invariable tendency of electricity is to
seek equilibrium. Where there is a plus of
force, the negative pole absorbs and carries
it off; where there is a minus, the positive
flows in to supply the deficiency, hence inflammatory surfaces require the application of
the negative pole; torpid or paralytic parts
need the quickening action of the positive.
Meantime, regarding the human frame as an
assemblage of magnetic and electric centres
in which force is generated and distributed
thoroughout the system on the principle of a
battery, the brain must be the grand central
positive pole; the spinal cord and extending

nerve fibres, the conducting wires, and the extremities, the negative polar points. The ganglionic system of nerves fulfills the same office for the viscera within the organism, as the cerebro-spinal system performs without. Besides the grand positive polar centres resident in the brain, there are two equally important positve vital centres in the lungs and heart: thus we may say in brief, and without embarrassing this simple little treatise with disquisitions on vital electricity, that we regard the brain, lungs and heart, as the chief positive centres of vital force in the body; hence also that we can never apply electricity safely to these points, because we should reverse the normal flow of the life currents, and instead of promoting the action which nature so clearly designs of a continuous and uninterrupted flow from positive to negative, from the brain, lungs, and heart, to their corresponding negative poles scattered throughout the entire system, we should cause a reversal of that natural order, and attract back the electric flow in injurious and congestive ex-

cess towards these important centres. We
may treat successfully diseases of these parts,
but not by direct application as will hereafter
be shown ; moreover, in imitation of nature's
sublime order, and in aid of her disturbed func-
tions, our applications of electricity should
always be made with *the positive pole applied
above the negative in position on the body.*
There are some few cases in which a reversal
of this rule is allowable, as in paralysis, where
it is essential to energize torpid or revive
dead nerves by an up-toning current ; we
may also, in cases of excessive debility, find
our advantage in directing an upward current
with a view of excessive stimulation ; also it
is frequently advisable to send a through current back and forth in such organs as we can
reach from opposite surfaces, as when electrodes or plates are applied to the kidneys
from the surfaces across the loins and abdomen ; but these exceptional applications will
be fully met in the directions herein laid
down, and our only duty is now to impress
upon our readers, as safe general rules, the

propriety of making no direct application of electricity to the head, lungs, or heart, nor yet of placing the positive pole below the negative in position on the body, except always in such methods as will be advised in this treatise.

Sec. V. The Position and Treatment of the Patient.

As intimated in Sec. IV., it is objectionable to make direct applications of electricity to the brain, lungs or heart ; yet diseases which immediately affect these great polar centres must be met, and can be most successfully treated with electricity, for example, in most conditions of headache, even in apoplexy and congestion of the brain, it is safe and curative to apply a small plate (Nos. 5 or 4) across the medulla oblongata, or base of the brain, positive and another negative at some expedient point below.

In asthma, bronchitis, and all affections of the lungs, it is safe and curative to apply plates Nos. 5 or 4 to the sternum or breast bone lengthways, or to rub the sponge down the length of the sternum.

For diseases of the heart no actual cure exists but by treating the system and restoring healthful vital action to the whole frame ; all cardiac difficulties can be modified and subdued even if they cannot be cured.

In treatments for toothache or facial neuralgia, an effective method of cure is to let the operator hold an electrode positive in the left hand, stroke the part affected gently with the right hand, and apply a plate negative to the base of the brain. If the operator is not in good health or objects to administer electricity through the hand, the sponge-cup attached to the positive may be used with good effect passed over the painful part. Directions also will be given for the treatment of the eye and ear by local applications, but these are almost the only exceptions we can make to the rules laid down in Sec. IV.

As for the position of the patient in receiv-

3

the treatment, the most favorable is the most comfortable, and the following arrangements will best meet all cases.

Let the operating room be pleasantly warm ; the patient, if a lady, should remove her outer garments, corsets, and all compressive bands, strings, &c. Gentlemen should do the same to the extent necessary.

In ordinary treatments the patient should then lie down on a couch if possible, continuing during the treatment on the back. This will give the operator an opportunity to place the plates on both the back and front of the organism as desired. Where plates are applied to the spine, back of the neck, under the calves of the legs, etc., small cushions or towels rolled up into a compress are found useful to press the plate against the body. Unless the plate touches the body in direct contact, no result will be obtained, and as there is often a hollow or space between the couch and some portion of the body on which plates may be required, the compresses advised above may be necessary. When it is desirable to place the feet in water, the patient may sit up in a high-backed chair, by aid of which a plate can be applied to the spine or between the shoulders. Should it be desired to use a plate at the base of the brain, together with the foot-tub, the plate may be bound on the neck with a handkerchief. In using electricity to the feet, the large plate, No. 1, may be employed, but this must be kept very warm and the feet covered up or placed near a fire. The foot-tub answers the purpose of retaining the heat better than a plate, in which case a metal tub must be employed, wooden pails or bowls carrying off the current too rapidly. Let the water be as warm as can be borne, and as the surface alone is electrified and it may be desirable to confine the force as much as possible to the soles of the feet, use only about an inch of water, increasing it to keep the heat up from time to time, and add as a general rule a handful of salt or a little saleratus as may be desired.

Let the patient be covered over with a spread in lying down, and well wrapped up in any position.

Where it becomes necessary to stroke down the spine with a sponge-cup, great care should be taken to avoid draughts of air, and at all times guard against the patient's catching cold.

Sec. VI.   On the Battery and its Connection with Electrodes.

Every electrical machine is sent out from the manufactory with certain concise directions for its use pasted in the cover, or added in the form of a phamphlet ; still it is desirable to find these directions in such a guide as this treatise claims to be, and therefore we may say in brief, that as the operator stands facing the battery, the posts or attachments for the strings will be found on the left hand, positive, on the right, negative.

There is also a cylinder or piston at the end of the box which increases or diminishes the strength as it is pulled out or pushed in, but as this arrangement varies in different machines, so it is necessary to consult carefully the directions pasted in the cover, before attempting to use the battery.

The strength of the current should be wholly regulated by the sensations of the patient.

So subtle are the varieties that pervade the human organism, that it is hardly probable to expect out of one hundred persons, that the current should affect any two exactly alike. The true gauge will be found therefore in the comfort of the patient, and the agreeable sensations the current produces. If it seems to clutch, draw, or prove weighty, the amount of force should be instantly reduced. *It should never cause pain or discomfort.* Where it does, the effect is invariably injurious. Also it must be remembered that though there is the same amount of force at both the positive and negative poles, there is a very different quantity of sensational action in different parts of the body, especially in diseased conditions. The plates at one point may produce the most lively and even painful sensations, while at another no sensation or very little may be manifest.

The finger ends being abundantly supplied with nerves, are often highly sensational, whilst another part of the organism may be almost insensible ; as a rule, therefore, it is better to regulate the current to the most sensational point. Thus, when the sensation is as strong as can be comfortably borne at one point, let this become the gauge, and do not increase the force to suit another, even if that other be a more important point of the organism. If it is deemed desirable to feel the force in some particular place, remove the second plate to some correspondingly expedient point, until ˉan agreeable sensation is realized ; for example, if it is desired to remove a severe headache, and a small plate is placed at the base of the brain, and another laid under the hands, with a view of drawing off inflammation from the head, supposing that the sensation is very strong in the hands, and none is felt at the neck at all, — in that case, it would be advisable to remove the plate from the feet to the calves of the legs, or some distant point from the head, and let the sensation become sufficiently vivid in the neck.

In connecting the battery it is always advisable to start the battery before touching the patient ; see that it runs smoothly, and then place the plates one after the other on the points desired, connecting the person with the battery, after it is started, by means of the pins attached to the strings. In this way the first shock of the connection will be made lighter after the battery is started, than when it is started after the plates and pins are fixed. The first entering in of the current is the strongest sensation produced, hence it is most desirable to commence with a very light current, and increase afterwards ; also in making changes always withdraw one of the pins before closing or making a change. Before recommencing after a change, diminish the strength, thus making the first entrance of the current, as above suggested, as light as possible. In this way the dangers and disagreeable sensations of sudden shocks will be avoided. Finally, we will add, first, that the conditions of every patient, and their capacity to take electricity comfortably, varies from day to day, and is continually influenced by changes in their own physical condition, the state of the atmosphere, and even their frame of mind or temper.

Always begin therefore with a light current, and in the manner described above. Also—and as one of the most important recommendations we can give—be careful not to exceed a séance of from 20 to 30 minutes in duration. To extend an electric séance beyond that time is to render the patient liable to reactionary fatigue, weariness, and pain. For merely local treatment, as for sore throat, headaches, scalds, burns, &c., ten or fifteen minutes may be quite sufficient at a time, but thirty minutes is the utmost duration that should be indulged in, always bearing in mind that too much electricity is an overdose of nature's life principle,—a drug which can cure as well as cause diseases, heal and save life as well as destroy it.

Sec. VII. Cephalagia, or Headache.

The varieties and characteristics of Cephalagia may be termed "legion," so numerous are the diseases of the various organs which ultimately represent themselves in headaches. To attempt dealing with this affection separately would scarcely be possible, as with the exception of the pain and weariness which proceeds directly from over-taxation of the brain, eyes, or ears, headaches may result from a disease of almost any organ of the body : for example, affections of the stomach, liver, spleen, uterus, rheumatism, neuralgia, over-exertion of the muscles in walking, riding, gymnastics, &c.,—all these, no less than fevers of every description, and even heart and lung diseases, are fruitful sources of Cephalagia. It is not necessary, then, to direct treatment for this painful and distressing affection alone : we must make up our minds in administering treatment for the headache, either to give electricity as a mere paliative or for special disability of which it is a symptom. For the present, however, we may classify certain forms of headaches under a few simple roots, and by prescribing for these we shall cover all the ground necessary to enter upon in this section. We commence with Sec. VIII.

Sec. VIII.   Sick Headache.

As the stomach and digestive organs are involved in this difficulty, apply a plate, P (*Positive*), across the loins, and another, N (*Negative*), across the abdomen, five minutes, then remove the back plate up between the shoulders lengthways, P, and the front across the diaphragm (*upper part of stomach*), N, five minutes. Then apply a plate to the nape of the neck, P, and place the feet in a metal foot-tub with about an inch in depth of hot salt and water ; into this drop the tin electrode N, and keep adding hot water to keep up the temperature for ten minutes. If this fails to relieve at once, the patient may have the narrow plate No. 5 placed on the nape of neck or between the shoulders, as is most agreeable, P holding the electrode N between the hands for ten minutes at a time, two or three times a day.

N. B. There are some persons in whom a naturally sensitive or congestive tendency of the head renders the application of a plate to the nape of the neck unpleasant and even injurious.

If there are any symptoms of fullness, increase of headache, or other disagreeable sensations, experienced by this application, substitute for it a plate lengthways between the shoulders.

Where the plate can be borne agreeably at the nape of the neck, it is an effective treatment, and one which produces good results in headache, sleeplessness, affections of the eyes, ears, catarrh, &c., but care must be taken to notice *the effects of this application ; and in whomsoever it causes any disagreeable sensations, the alteration above suggested must invariably be made. We may also observe here, that where it is difficult to procure hot water, or a metal foot-tub, the large plate No. 1 may be placed under the feet which, if kept well warmed, will act as effectively as the foot-tub.

Sec. IX.   Congestive Headaches.

Treat generally according to the directions given in Sec. VIII. ; but as congestive headaches are likely to be aggravated rather than assuaged by the application of a plate to the base of the brain, it may be better in all such cases to apply it as above directed between the shoulders.

Sec. X.   Neuralgic Headaches.

If it can be borne, a plate, P, to the nape of the neck and electrode in hands, N, ten minutes ; repeat as often as relief is obtained.

Sometimes it is an effective treatment for the patient to hold the electrode N in the hands whilst the operator applies the sponge-cup P to the nape of neck, rubbing the sponge upwards into the hair at the back of the head.

Sec. XI. Nervous Headaches.

As all forms of nervous derangement affect the system, and sometimes proceed from it exclusively, it is well in nervous headaches to tone up the patient by the "systemic treatment" (Sec. XIV.). We need but add here, if the patient desires to take treatment for the head alone, place an electrode, N, in the hands, and brush up from the nape of the neck with the sponge-cup P, also rest the sponge on the neck ; in all, ten minutes.

Sec. XII.   For Headaches proceeding from Uterine Weakness, or accompanied with Debility and Pain in the Back.

Let the patient sit on the large plate No. 1 well warmed, N, and apply a plate, P, to the nape of the neck or between the shoulders, as can be best borne, ten minutes.

Sec. XIII.   Systemic Headaches Generally.

Sit on the large plate N, and stroke down the whole length of the spine with the sponge-cup, ten minutes; then sit on the large plate P, and place the feet, N, in a foot-tub with hot salt and water, or on another large plate well warmed, ten minutes.

N. B.   The above direction will prove especially useful to females for hastening the monthly flow or increasing it when sluggish.

Sec. XIV. General Systemic Treatment for Debility, Consumption, Dyspepsia, &c., in Males or Females.

Plate across loins, on lowest part of spine, P, and then plate across lowest part of abdomen, five minutes ; then reverse the current, by changing the pins, five minutes ; then move up the back plates to between the shoulders lengthways, P, and place abdomen plate across diaphragm, N, five minutes ; then reverse the current, by changing the pins, five minutes ; then a small plate on nape of neck, P, and another under the fingers laid flat on the plate, N, six minutes.

Sec. XV. For Sore Throat.

Small plate at back of neck, P, and the sponge-cup rubbed gently over the front of the throat, N, from six to ten minutes. In case of ulceration, swelled tonsils, &c., the operator may use the spatula (if possessed of instruments), and apply it, covered with a fine cloth, so as gently to touch the swellings for a minute or two at a time. In Diptheria, Bronchites, or Hoarseness, the sponge-cup rubbed over the front of the throat, or an application of electricity through the operator's hand, can be repeated several times a day during from five to ten minutes at a time.

Sec. XVI. For Croup, Asthma, or Constriction of the Chest.

A plate at nape of neck, P, and the sponge-cup rubbed gently down over the part most affected, N. In asthma, rub the sponge-cup well down the sternum, also remove the plate from the nape of the neck, and apply it between the shoulders, rubbing down the sternum (breast-bone) for ten minutes at a time. This treatment is generally good for all pulmonary affections, but in regular consumption the systemic treatment (Sec. XIV.) is advisable.

SEC. XVII. FOR LIVER COMPLAINTS.

Consult Sec. XIV., and apply the plates as directed for the first ten minutes, then, if the liver is inflamed and the surfaces tender, apply a plate, P, between the shoulders lengthways, and a small plate above, below and around the tender points, N, for two minutes at a time; finish with a small plate, P, at nape of neck and another, N, under hands.

N. B. When any organs or parts prove painful under the plates, when there is great tenderness or a sense of extreme weight, remove the plate to a point near it, avoiding only the surfaces over the heart or lungs; also when a reverse current is directed, and it affects the patient unpleasantly, or produces a sense of fullness, return at once to the normal current.

4

Sec. XVIII. For all Affections of the Kidneys.

The systemic treatment No. 14 is recommended ; only in the second part, or second change, apply one plate across the kidneys, P, and another between the shoulders at the same time, N. Another method of relief, and one very effective where there is much pain and inflammation of the kidneys, is the treatment at Sec. XIII. The same treatment (XIII.) is effective in diseases of the bladder. If there is any inflammation of the bladder, place a plate across the loins, P, and one on the abdomen, N, or rub over the lower part of the abdomen for ten minutes with the sponge-cup. Then sit on a large plate, P, and apply a foot-tub or plate to the feet, N, ten minutes.

Sec. XIX. For Sciatica.

Place a plate, N, under the thigh on the side affected, and rub over the painful part well with the sponge-cup, P ; rub also over the loins and across the hip ; end with sitting on a large plate, P, and another or a foot-tub, N, at the feet.

Sec. XX. Paralysis.

If one side only is attacked, place a narrow plate between the shoulders, N, and rub over the parts affected with sponge-cup, P. If the paralysis affect the spine, place one plate between the shoulders lengthways, and another across loins (low down), and reverse the current frequently, sending a current up and down. If one or both extremities are paralyzed, place the foot or feet, hand or hands, in strong saleratus water (hot), and plates on the spine between the shoulders at the nape of the neck, and across the loins, three minutes each, or, what is better still, rub down the length of the spine with a sponge-cup. In all cases of paralysis use alternately the normal or down current, and the up, or toning current, also fail not to treat the system occasionally as at Sec. XIV.

Sec. XXI. Paralysis of the Face, Tongue, or Cranial Nerves.

In this case it is allowable to rub the sponge-cup, P, over the parts affected whilst an electrode is held, N, in the hands, or the feet are immersed, N, in foot-tub. If the speech is affected, rub the sponge-cup on the nape of the neck and over the front of the throat, P.

Sec. XXII. For Ophthalmia, or Sore Eyes.

Place a plate, N, between shoulders, and hold the eye-cup, filled half with water and half with witch-hazel, rose-water, or weak solution of alum water, close to the eye, so close indeed that the lashes are steeped in the fluid; let this be applied to the eye, P, with a gentle current, in fact only sufficient to cause a slight tingling sensation for five minutes at a time, to one or both eyes as may be required.

In AMAUROSIS, the plate between the shoulders must be P, the eye-cup, N. In case no eye-cup is accessible, let the operator lay the two fore-fingers of the right hand gently on the closed lid, holding an electrode in the left hand, or, if objectionable to pass electricity through the operator's hand, use a sponge-cup steeped in the solution directed above.

Use also a good but mild eye-wash from time to time, each day and night.

N. B. Let the electricity be felt, but the current be exceedingly mild in treating the eyes.

SEC. XXIII. FOR PARALYSIS OF THE AUDITORY NERVE, AND DEAFNESS.

It will sometimes suffice to improve, if not entirely to restore the hearing, to apply a small plate or the sponge-cup, P, to the nape of the neck ten minutes at a time each day, and let the patient hold an electrode, N, in the hands.

Where this treatment fails, the sponge-cup may be rubbed around the ear on the outside, and a small ear-instrument, tipped with a piece of fine sponge moistened, be applied for for a minute or two at a time inside the ear.

The first treatment recommended should be tried well before the second.

Sec. XXIV.   For Chorea, or St. Vitus Dance.

Sit on a large plate, N, and stroke down the back, P, five to eight minutes; then place a large plate beneath the feet, N, and make the seat-plate, P, five minutes; end with a small plate at nape of neck, P, and another, N, under the calves of legs, five minutes.

Sec. XXV.   For Rheumatism.

If of the extremities, place the foot or hand in hot saleratus and water, N, and bind a plate above the painful part, P, or rub over its surface downwards with sponge-cup ten minutes. If of the shoulder or breast, place plates lengthways between the shoulders and up and down the breast-bone. If of the neck, rub over the parts with the sponge-cup, P, and hold an electrode in the hands, N. For gout, use the first treatment directed above, and the systemic treatment to renovate the system.

Sec. XXVI. For Cutaneous Diseases, and Rheumatism of the System.

Use, if possible and compatible, a warm saleratus bath, in which an electrode, N, is dropped, whilst the sponge-cup, P, is applied either to any specially painful part, or held in the hands. The bath should not be taken more than from fifteen to twenty miuutes.

Sec. XXVII. In Influenza, Severe Cold, Fever and Ague.

Use the systemic treatment, Sec. XIV. In Fever and Ague, in especial, take the treatment in advance of the expected attack.

Sec. XXVIII.   For Diarrhœa or Dysentery.

Wrap a tin electrode round with a thin piece of sponge ; tie this and use it, P, placing it against the entrance to the rectum. Sit on this, and lay a large, very hot plate, N, over the bowels ten minutes. Then keep the plate over the bowels, P, and place another opposite to it across the lower part of the back, N, for ten minutes.

Sec. XXIX.   For Constipation.

Place the electrode and sponge, as directed in Sec. XXVIII., at the rectum, and knead over the bowels thoroughly with the sponge-cup for ten minutes. Repeat this treatment, with intervals of an hour, until relieved.

Sec. XXX.   For Bruises, Burns, Scalds, Gatherings, or Boils.

If on the extremities, place the hand or foot in warm water with an electrode, N, and bind a plate, P, above the part affected ten minutes. If the seat of injury is on the body, lay a warm plate, N, over it, with two or three folds of damp cloth intervening, and send a current through the body to the opposite point, P.

Use the same treatment for sprains or inflamed joints, always remembering that the negative cools, the positive warms.

Sec. XXXI. For Colic, Cramp in Stomach or Bowels.

The through treatment as in Sec. XIV., using large plates, and placing hot cloths over the plates to keep in and promote excessive warmth. Use till relieved.

Sec. XXXII. For Piles or Hemorrhoids.

Sit on electrode, wrapped round with sponge, N, and rub down the spine with sponge-cup, P, ten minutes; then use the through treatment, as in Sec. XIV., first ten minutes.

XXXIII. In Cases of Small Pox.

Chicken Pox, Whooping Cough, or children's diseases generally, electricity is made an effective treatment by placing narrow plates on the breast-bone and spine, or on the spine and stomach, for about ten minutes at a time. It will also soothe restlessness and promote sleep by an electrode being held in the hands, and a small plate applied to the base of the brain for from six to ten minutes at a time. This is also an effective remedy in fevers, and the sponge-cup can be safely used at the throat, or down the breast-bone, when either part is particularly affected.

Let the student who would treat any case successfully, not only observe the directions herein specially laid down, but thoroughly master all the preliminary details with respect to the position and treatment of the patient, the management of the battery, strength of the current, and other matters, no one of which is of subordinate importance to the other. Let the patient be kept always warm and comfortable, or if too hot and feverish, cool and airy. Never give the current too strong or too long ; never connect the battery suddenly : always keep the positive above the negative, except when directed otherwise ; and when all these, and every other detail laid down in the earlier pages of this treatise, are fully mastered, the operator can apply the general principles to a much wider range of complaints than are herein laid down.

We cannot conscientiously advise self-treatment, or recommend domestic practice, in severe, acute, dangerous, or surgical cases. These excepted, the student who faithfully masters and carefully practices the directions contained in this guide, need never be at a loss for a means of cure, or resort to a physician's aid.